Discrete and Continuous Probability Distributions: A Creative Comparison

By Kathryn Paulk © 2018

Updated: 08/01/2025

TABLE OF CONTENTS:

PREFACE 5

INTRODUCTION 6

DISCRETE PROBABILITY DISTRIBUTIONS 7
    DISCRETE -- PROBABILITY MASS FUNCTION (PMF) 8
    DISCRETE -- CUMULATIVE DISTRIBUTION FUNCTION (CDF) 10
    DISCRETE -- EXPECTED VALUES AND BASIC STATISTICS 11
    DISCRETE -- PROBABILITY OF (X = 3) 13
    DISCRETE -- PROBABILITY OF (2 <= X <= 4) 14
    DISCRETE -- PROBABILITY OF (3 <= X) 15
    DISCRETE DISTRIBUTION SUMMARY 16

CONTINUOUS PROBABILITY DISTRIBUTIONS 17
    CONTINUOUS -- PROBABILITY DISTRIBUTION FUNCTION (PDF) 18
    CONTINUOUS -- CUMULATIVE DISTRIBUTION FUNCTION (CDF) 21
    CONTINUOUS -- EXPECTED VALUES AND BASIC STATISTICS 22
    CONTINUOUS -- PROBABILITY OF (X = 3) 23
    CONTINUOUS -- PROBABILITY OF (2 <= X <= 4) 24
    CONTINUOUS -- PROBABILITY OF (3 <= X) 25
    CONTINUOUS -- PROBABILITY OF (X > 5) 26
    CONTINUOUS DISTRIBUTION SUMMARY 27

NORMAL DISTRIBUTION -- ADDITIONAL INFORMATION 28
    FIND PERCENTILE FOR GIVEN OVERTIME 28
    FIND OVERTIME FOR GIVEN PERCENTILE 29
    COMPARE ACTUAL DATA WITH NORMAL DISTRIBUTION 30
    CREATE A NORMAL PROBABILITY PLOT 32
    CREATE AN "AB-NORMAL" PROBABILITY PLOT 34

MORE CONTINUOUS PROBABILITY DISTRIBUTIONS 35
    BACKGROUND INFORMATION – GAMMA DISTRIBUTION 36
    MORE CONTINUOUS PROBABILITY DISTRIBUTIONS (PMF) 37
    CUMULATIVE PROBABILITY (CDF) 38
    EXPECTED VALUES AND BASIC STATISTICS 39
    PROBABILITY OF (X = 3) 40
    PROBABILITY OF (2 <= X <= 4) 41

INCENTIVES **43**
MONEY INCENTIVES - DESCRIPTION 43
MONEY INCENTIVE CALCULATIONS – DISCRETE 44
Discrete Incentive Calculations – Mom 44
Discrete Incentive Calculations – Dad 46
Discrete Incentive Calculations – Aunt 48
Discrete Incentive Calculations – Uncle 49
Discrete Incentive -- Summary 52
MONEY INCENTIVE CALCULATIONS – CONTINUOUS 53
Continuous Incentive Calculations – Mom 53
Continuous Incentive Calculations – Dad 54
Continuous Incentive Calculations – Aunt 55
Continuous Incentive Calculations – Uncle 56
Continuous Incentive -- Summary 57

APPENDIX **58**
THE ORIGINAL PROBLEM 59
Problem Statement 59
Problem Solution 60
THE OPTIMIST'S CREED 62

OTHER BOOKS BY KATHRYN PAULK **63**

Preface

Statistics is a powerful tool, if you know how to apply it properly. To use statistics, one must identify the type of distribution that models a particular situation. Then, the equations must be correctly applied. It is assumed that the reader has completed courses in differential and integral calculus.

To help students with identifying statistical distributions and applying appropriate equations, I personified a set of discrete and continuous distributions with a set of college professors, who stay overtime in their classes, according to a particular distribution. Statistics, for each professor, are calculated and compared.

I would like to express my gratitude to Dr. Yuly Koshevnik, a Senior Lecturer at the University of Texas at Dallas, who reviewed this document and provided many helpful comments and suggestions. In particular, he recommended adding the geometric distribution to the set of discrete distributions, discussed in this book.

The inspiration for this detailed comparison of discrete and continuous distributions came from a question in the book: <u>Probability & Statistics for Engineering and the Sciences</u>, Eighth Edition, Jay L. Devore, Chapter 4, p. 142, question #5. The problem statement and solution are included in the appendix. The problem was expanded to include both discrete and continuous distributions and to answer additional questions.

Copyright © Kathryn Paulk, 2018

Introduction

The purpose of this document is to enhance the student's understanding of probability distributions by comparing some discrete and continuous distributions with one simple, silly, and relatable example.

The distributions are represented by different professors who run their classes overtime, according to discrete or continuous distributions. The discrete and continuous distributions, included in this example, are listed in the following table.

Discrete Distributions	- Discrete
	- Binominal
	- Poisson
	- Geometric
Continuous Distributions	- Uniform
	- Normal
	- Exponential
Additional Continuous Distributions	- Standard Gamma
	- Chi-Squared (χ^2)

Discrete Probability Distributions

Discrete Distributions	• Discrete
	• Binominal
	• Poisson
	• Geometric

Discrete -- Probability Mass Function (pmf)

Meet the professors who represent discrete distributions. Their middle name is "Discrete." They all hold their classes overtime, in their own way.

Professor	Discrete -- Probability Mass Function (pmf)
Professor D. Discrete	For the past few years, students have been collecting data on the overtime for this guy. His overtime is always between 1 and 5 minutes with this probability distribution. | X | 1 | 2 | 3 | 4 | 5 | | P(x) | .50 | .10 | .20 | .05 | .15 |
Professor D. Binomial	This guy will roll <u>five</u> 6-sided dice. The number of 1's = # minutes of overtime. Here: $n = 5, \quad p = \frac{1}{6} = .1667$ When none of the six die has a "1" then there is no overtime! So, actually, his range is from 0 to 5 minutes of overtime. $b(x; n, p) = \binom{n}{x} p^x (1-p)^{n-x}$ $b\left(x; 5, \frac{1}{6}\right) = \binom{5}{x} (.1667)^x (.8333)^{5-x}$
Professor D. Poisson	On the average, this guy will stay overtime about 1 hour during the term which includes 30 classes. Here, on the average: $\mu = \frac{1\ hour}{30\ classes} = \frac{60\ mins}{30\ classes} = \frac{2\ mins}{class} = 2$ $p(x, \mu) = \frac{e^{-\mu} \mu^x}{x!}$ $p(x, 2) = \frac{e^{-2} 2^x}{x!}$

Professor	Discrete -- Probability Mass Function (pmf)
Professor D. Geometric	This guy is a pessimistic person. He's the kind of person who is never happy with what he has and always focuses on what he does not have. At the end of the class, he will roll one 6-sided die until he rolls a "1." He counts the failures. When he has success (rolls a "1") he stops counting. The number of failures "F" are the number of minutes of overtime he stays. $$P(X = x) = (1 - p)^x p$$ Number of failures before success $= x$ The probability of success is: $p = \frac{1}{6}$ The expected number of failures (before 1 success) is given by: $$E(X) = \frac{(1-p)}{p} = \frac{\left(\frac{5}{6}\right)}{\left(\frac{1}{6}\right)} = \left(\frac{5}{6}\right)\left(\frac{6}{1}\right) = \frac{30}{6} = 5$$ Since this guy is so pessimistic and students can expect to stay 5 minutes of overtime for every class, most students avoid taking his classes.

Discrete -- Cumulative Distribution Function (cdf)

Professor	Discrete -- Cumulative Distribution Function (cdf)
Professor D. Discrete	<table><tr><td>x</td><td>1</td><td>2</td><td>3</td><td>4</td><td>5</td></tr><tr><td>cdf</td><td>.50</td><td>.60</td><td>.80</td><td>.85</td><td>1</td></tr></table>
Professor D. Binomial	$B(x; n, p) = P(X \le x) = \sum b(x; n, p)$ $B(x; 5, .1667) = $ Use Binomial Tables!! Use Tables! $p = .17$ (between .10 & .20)
Professor D. Poisson	$F(x, \mu) = P(X \le x) = \sum \frac{e^{-\mu} \mu^x}{x!}$ $F(x, 2) = \sum \frac{e^{-2} 2^x}{x!}$ (Tables are much easier!!)
Professor D. Geometric	$P(X \le x) = F(p, x) = 1 - (1-p)^{(x+1)}$ $x = $ # failures before first success $p = $ probability of success Use some common sense to check this: • Suppose a "1" is rolled the first time. Then $x = 0$. • $P(X \le 0) = 1 - (1-p)^{(0+1)} = 1 - \left(\frac{5}{6}\right) = \frac{1}{6}$ • This probability should be the same as the probability of rolling a "1" on the first trial. And, it is!

Discrete -- Expected Values and Basic Statistics

Professor	What value would you expect?	What are the basic statistics? (mean, variance, and standard deviation)
Professor D. Discrete	$E(X)$ $= (.50)1 + (.10)2$ $+ (.20)3 + (.04)4$ $+ (.15)5 = 2.21$	$mean = \mu = E(X) = 2.21$ 2 ways to calculate variance: $V(X) = \sum(x - \mu)^2 p(x)$ $V(X) = E(X^2) - [E(X)]^2$ $E(X^2)$ $= 1(.50) + 4(.10) + 9(.20)$ $+ 16(.05) + 25(.15) = 7.25$ $V(X) = E(X^2) - [E(X)]^2$ $= 7.25 - (2.21)^2 = 2.366$ $\sigma = \sqrt{V(X)} = \sqrt{2.366} = 1.54$
Professor D. Binomial	$E(X) = np$ $E(X) = (5)(.1667)$ $= .8333$	$mean = \mu = E(X) = .8333$ $V(X) = npq \; ; \; q = (1 - p)$ $V(X) = 5(.1667)(.8333) = .6945$ $\sigma = \sqrt{V(X)} = \sqrt{.6945} = .833$
Professor D. Poisson	$E(X) = \mu$ $E(X) = 2$	$mean = \mu = E(X) = 2$ $V(X) = E(X) = 2$ $\sigma = \sqrt{V(X)} = \sqrt{2} = 1.414$

Professor	What value would you expect?	What are the basic statistics? (mean, variance, and standard deviation)
Professor D. Geometric	$E(X) = \frac{(1-p)}{p}$ $= \frac{\left(\frac{5}{6}\right)}{\left(\frac{1}{6}\right)} = \left(\frac{5}{6}\right)\left(\frac{6}{1}\right)$ $= \frac{30}{6} = 5$ Probability of success is the probability of rolling a "1" → $p = \frac{1}{6}$	$mean = \mu = E(X) = 5$ $V(X) = \frac{1-p}{p^2} = \frac{\left(\frac{5}{6}\right)}{\left(\frac{1}{6}\right)^2} = 30$ $\sigma = \sqrt{V(X)} = \sqrt{30} = 5.477$

Discrete -- Probability of (X = 3)

Professor	Probability of staying exactly 3 minutes overtime.
Professor D. Discrete	$P(X = 3) = .20$ Another Way (Using your cdf Table)... $P(X = 3) = P(X \leq 3) - P(X \leq 2) = .80 - .60 = .20$
Professor D. Binomial	$P(X = 3) = b(3; 5, .1667)$ $\quad = \binom{5}{3}(.1667)^3(.8333)^{5-3}$ $\quad = (10)(.0043)(.694) = .0299$ Another Way (Using the Binomial Tables) ... $P(X = 3) = P(X \leq 3) - P(X \leq 2)$ $\quad = B(3; 5, .1667) - B(2; 5, .1667)$ $\quad = B(3; 5, .17) - B(2; 5, .17) = .9965 - .9665 = .03$
Professor D. Poisson	$P(X = 3) = p(x, 2)$ $\quad = \dfrac{e^{-2} 2^3}{3!} = \dfrac{1.0827}{6} = .18045$ Another Way (Using the Poisson Tables) ... $P(X = 3) = P(X \leq 3) - P(X \leq 2)$ $\quad = F(3,2) - F(2,2) = .857 - .677 = .18$
Professor D. Geometric	$P(X = x) = (1 - p)^x p$ $P(X = 3) = \left(\dfrac{5}{6}\right)^3 \left(\dfrac{1}{6}\right) = .0965$ NOTE: The probability of 3 failures before 1 success is like the probability of not rolling a "1" three times in a row, followed by the probability of rolling a "1".

Discrete -- Probability of (2 <= X <= 4)

Professor	What's the probability of staying between 2 and 4 min. Inclusive?
Professor D. Discrete	$P(2 \leq X \leq 4) = P(2) + P(3) + P(4)$ $= .10 + .20 + .05 = .35$ Another Way (Using cdf Table) $P(2 \leq X \leq 4) = P(X \leq 4) - P(X \leq 1) = .85 - .50 = .35$
Professor D. Binomial	$P(2 \leq X \leq 4) = P(2) + P(3) + P(4)$ $= b(2; 5, .17) + b(3; 5, .17) + b(4; 5, .17)$ $= A\ lot\ of\ work!!!$ ➔ Use Binomial Table $P(2 \leq X \leq 4) = P(X \leq 4) - P(X \leq 1)$ $= B(4; 5, .17) - B(1; 5, .17)$ $= 1 - .828 = .172$
Professor D. Poisson	$P(2 \leq X \leq 4) = P(2) + P(3) + P(4)$ $= p(2; 2) + p(3; 2) + p(4; 2) = A\ lot\ of\ work!!!$ Much Easier to use Poisson Table… $P(2 \leq X \leq 4) = P(X \leq 4) - P(X \leq 1)$ $= F(4; 2) - F(1; 2) = .947 - .406 = .541$
Professor D. Geometric	$P(2 \leq X \leq 4) = P(2) + P(3) + P(4)$ $= \left(\frac{5}{6}\right)^2 \left(\frac{1}{6}\right) + \left(\frac{5}{6}\right)^3 \left(\frac{1}{6}\right) + \left(\frac{5}{6}\right)^4 \left(\frac{1}{6}\right)$ $= .116 + .097 + .080 = .293$

Discrete -- Probability of (3 <= X)

Professor	What's the probability of staying at least 3 minutes overtime?
Professor D. Discrete	$P(at\ least\ 3\ mins) = 1 - P(less\ than\ 3\ mins.)$ $P(X \geq 3) = 1 - P(X \leq 2)$ $\ldots\ = 1 - [P(X = 2) + P(X = 1)] = 1 - [.10 + .50] = .40$ Another Solution (Using cdf Table) $P(X \geq 3) = 1 - P(X \leq 2) = 1 - .60 = .40$
Professor D. Binomial	$P(at\ least\ 3\ mins) = 1 - P(less\ than\ 3\ mins.)$ $P(X \geq 3) = 1 - P(X \leq 2) = 1 - [P(X = 2) + P(X = 1)]$ $= 1 - [b(2; 5; .17) + b(1; 5, .17)] = A\ lot\ of\ work!!!$ Much Easier to use Binomial Table... $(X \geq 3) = 1 - P(X \leq 2) = 1 - B(2; 5, .17) = 1 - .967 = .034$
Professor D. Poisson	$P(at\ least\ 3\ mins) = 1 - P(less\ than\ 3\ mins.)$ $P(X \geq 3) = 1 - P(X \leq 2) = 1 - [P(X = 2) + P(X = 1)]$ $= 1 - [p(2; 2) + p(1; 2)] = A\ lot\ of\ work!!!$ Much Easier to use Poisson Table... $(X \geq 3) = 1 - P(X \leq 2) = 1 - F(2; 2) = 1 - .677 = .323$
Professor D. Geometric	$P(at\ least\ 3\ mins) = 1 - [P(2) + P(1) + P(0)]$ $P(X \geq 3) = 1 - \left[\left(\frac{5}{6}\right)^2 \left(\frac{1}{6}\right) + \left(\frac{5}{6}\right)^1 \left(\frac{1}{6}\right) + \left(\frac{5}{6}\right)^0 \left(\frac{1}{6}\right) \right]$ $= 1 - [.116 + .139 + .167] = 1 - [.422] = .578$ Also: $P(X \geq 3) = 1 - P(X \leq 2) = 1 - \left[1 - \left(\frac{5}{6}\right)^3 \right] = .578$ Also: $P(X \geq 3) = (1 - p)^3 = \left(\frac{5}{6}\right)^3 = .578$

Discrete Distribution Summary

Probabilities of Overtime (mins./class)	Professor Discrete	Professor Binomial	Professor Poisson	Professor Geometric
Expected Overtime	2.21	0.83	2	5
Probability of $(X = 3)$	.20	.03	.18	.12
Probability of $(2 \leq X \leq 4)$	.35	.17	.54	.24
Probability of $(3 \leq X)$	.40	.03	.32	.65

To minimize staying overtime for a class, for discrete distributions, take Professor Binomial's class.

Continuous Probability Distributions

Continuous Distributions	• Uniform
	• Normal
	• Exponential

Continuous -- Probability Distribution Function (pdf)

Meet the professors who represent continuous distributions. Their middle name is "Continuous." They all hold their classes overtime, in their own way..

Professor	Continuous -- Probability Distribution Function (pdf)
Professor C. Uniform	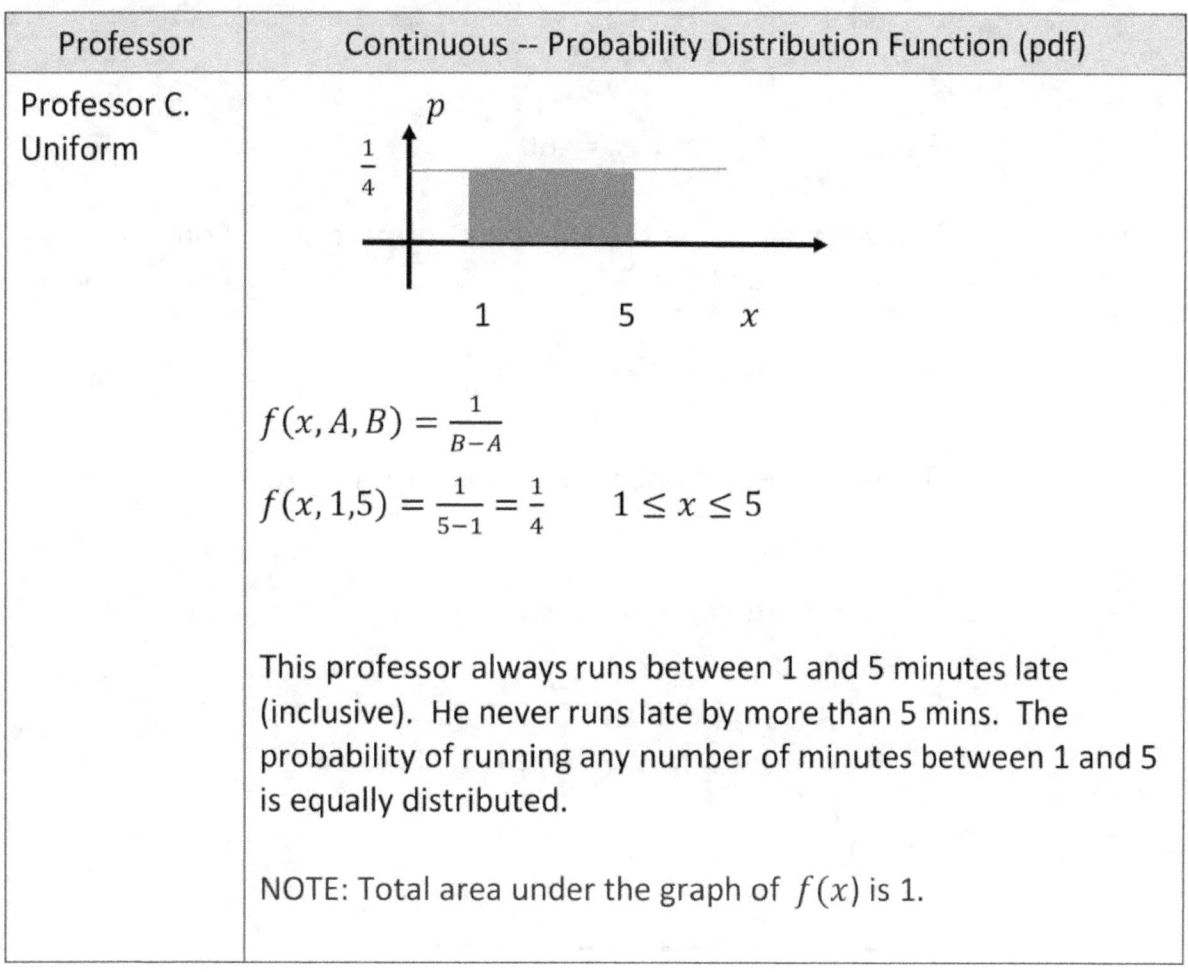 $f(x, A, B) = \frac{1}{B-A}$ $f(x, 1, 5) = \frac{1}{5-1} = \frac{1}{4}$ $\quad 1 \leq x \leq 5$ This professor always runs between 1 and 5 minutes late (inclusive). He never runs late by more than 5 mins. The probability of running any number of minutes between 1 and 5 is equally distributed. NOTE: Total area under the graph of $f(x)$ is 1.

Professor	Continuous -- Probability Distribution Function (pdf)
Professor C. Normal	*Graph of normal distribution with p on vertical axis and x on horizontal axis showing values 1, 3, 5*

Here: $\mu = 3$ and $\sigma = .60$

But we want to use the Standard Normal distribution with $\mu = 0$ and $\sigma = 1$

$$f(z; 0,1) = \frac{1}{\sqrt{2\pi}} e^{-z^2/2}$$

To use this equation, we need to convert x to z where $\mu = 3$ and $\sigma = .60$.

$z = \#\ standard\ deviations$
$z = \frac{x - \mu}{\sigma}$ ➔ $z = \frac{x - 3}{.60}$

X between 1 & 5 ➔ z between -3.33 & 3.33

Graph of standard normal distribution with p on vertical axis and z on horizontal axis showing values -3.33, 0, 3.33

This professor is usually 3 minutes late +/- a few minutes. He rarely runs 1 or 5 minutes late. |

Professor	Continuous -- Probability Distribution Function (pdf)
Professor C. Exponential	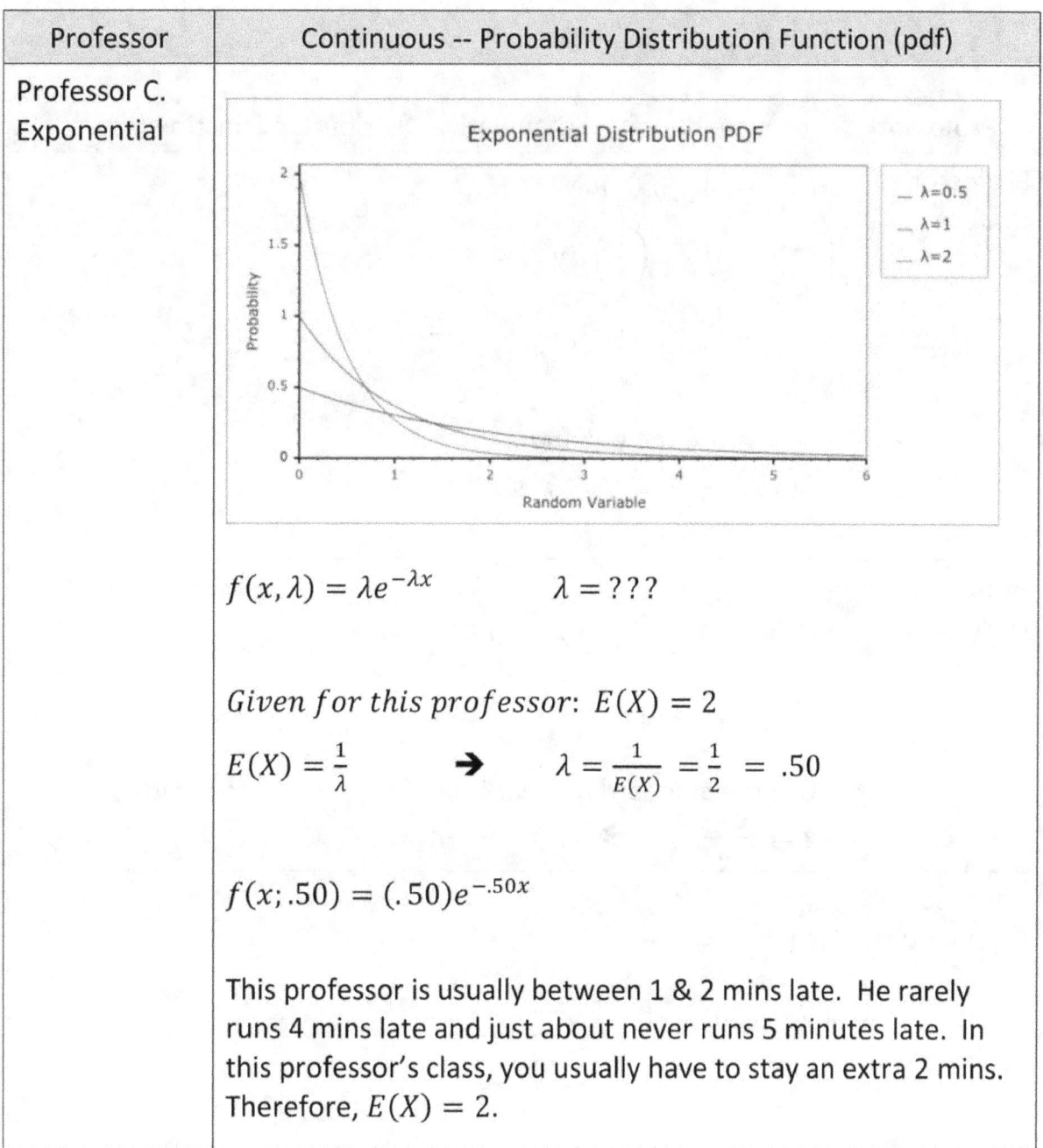 $f(x, \lambda) = \lambda e^{-\lambda x} \qquad \lambda = ???$ *Given for this professor:* $E(X) = 2$ $E(X) = \frac{1}{\lambda} \quad \rightarrow \quad \lambda = \frac{1}{E(X)} = \frac{1}{2} = .50$ $f(x; .50) = (.50)e^{-.50x}$ This professor is usually between 1 & 2 mins late. He rarely runs 4 mins late and just about never runs 5 minutes late. In this professor's class, you usually have to stay an extra 2 mins. Therefore, $E(X) = 2$.

Continuous -- Cumulative Distribution Function (cdf)

Professor	Continuous -- Cumulative Distribution Function (cdf)
Professor C. Uniform	$F(x) = \int_{-\infty}^{\infty} f(x)dx$ $= \int_{y=1}^{y=x} \left(\frac{1}{4}\right) dy$ $= \left[\frac{y}{4}\right]_1^x = \frac{x}{4} - \frac{1}{4} = \frac{x-1}{4} \qquad 1 \leq x \leq 5$ Or just use the equation: $F(x) = \frac{x-A}{B-A} \qquad A \leq x \leq B$
Professor C. Normal	$P(Z \leq z) = \int_{-\infty}^{z} f(y; 0,1) dy$ $P(Z \leq z) = \Phi(z)$ Use the Standard Normal Tables. Convert x to z with: $z = \frac{x-\mu}{\sigma} \qquad \rightarrow \qquad z = \frac{x-3}{.60}$
Professor C. Exponential	$F(x; \lambda) = 1 - e^{-\lambda x}$ $F(x; .50) = 1 - e^{-.50x}$

Continuous -- Expected Values and Basic Statistics

Professor	What value would you expect?	What are the basic statistics? (mean, variance, and standard deviation)
Professor C. Uniform	$E(X) = \frac{5+1}{2} = 3$	$mean = \mu = E(X) = 3$ $V(X) = \int (x-\mu)^2 \, p(x) \, dx$ $V(X) = \int (x-3)^2 \left(\frac{1}{4}\right) dx$ $V(X) = \left(\frac{1}{4}\right) \int_1^5 (x-3)^2 \, dx$ $V(X) = \left(\frac{1}{4}\right)(5.333) = 1.333$ $\sigma = \sqrt{V(X)} = \sqrt{1.333} = 1.155$
Professor C. Normal	$E(X) = \mu = 3$ Another way to calculate it: $z = \frac{x-3}{.60}$ → $x = (.60)z + 3$ $z = 0$ → $x = (.60)(0) + 3 = 3$	$mean = \mu = E(X) = 3$ Originally, we assumed $\sigma = .60$ So, $V(X) = \sigma^2 = .60^2 = .36$
Professor C. Exponential	$E(X) = \int_0^\infty x \cdot f(x) dx$ $= \int_0^\infty x(.50)e^{-.50x} dx = 2$ Easier way to calculate it: $\mu = \frac{1}{\lambda} = \frac{1}{.50} = 2$	$mean = \mu = \frac{1}{\lambda} = E(X) = 2$ $V(X) = \frac{1}{\lambda^2} = \frac{1}{2^2} = .25$ $\sigma^2 = V(X) = \frac{1}{\lambda^2}$ $\sigma = \sqrt{V(X)} = \sqrt{.25} = .5$

Continuous -- Probability of (X = 3)

Professor	Probability of staying exactly 3 minutes overtime.
Professor C. Uniform	$P(X = 3) = 0$ Can NOT calculate probabilities for a single value. Can only calculate probabilities for a range of x.
Professor C. Normal	$P(X = 3) = 0$ Can NOT calculate probabilities for a single value. Can only calculate probabilities for a range of x.
Professor C. Exponential	$P(X = 3) = 0$ Can NOT calculate probabilities for a single value. Can only calculate probabilities for a range of x.

NOTE: For ANY continuous distribution, the probability of a of particular value k is zero. In other words: For any value of k, $P(X = k) = 0$ for any continuous distribution. Continuous distributions are only defined for a range of values.

Continuous -- Probability of (2 <= X <= 4)

Professor	What's the probability of staying between 2 and 4 min. Inclusive?
Professor C. Uniform	$P(2 \leq X \leq 4) = ???$ $P(2 \leq X \leq 4) = F(4) - F(2)$ $\qquad = \left(\frac{4-1}{4}\right) - \left(\frac{2-1}{4}\right)$ $\qquad = \frac{3}{4} - \frac{1}{4} = \frac{2}{4} = .50$
Professor C. Normal	$P(2 \leq X \leq 4) = ???$ $x = 4 \rightarrow z = \frac{4-3}{.60} = 1.6667$ $x = 2 \rightarrow z = \frac{2-3}{.60} = -1.6667$ $P(2 \leq X \leq 4) = \Phi(1.667) - \Phi(-1.667)$ $\qquad = .9525 - .0475 = .905$
Professor C. Exponential	$P(2 \leq X \leq 4) = ???$ $F(x; .5) = 1 - e^{-.5x}$ $F(4; .5) = 1 - e^{-.5 \cdot 4} = .8647$ $F(2; .5) = 1 - e^{-.5 \cdot 2} = .6321$ $P(2 \leq X \leq 4) = F(4) - F(2)$ $\qquad = .8647 - .6321 = .2326$

Continuous -- Probability of (3 <= X)

Professor	What's the probability of staying at least 3 minutes overtime?
Professor C. Uniform	$P(at\ least\ 3\ mins) = 1 - P(less\ than\ 3\ mins.)$ $P(X \geq 3) = 1 - F(3)$ $\quad = 1 - \left(\frac{3-1}{4}\right) = 1 - \left(\frac{1}{2}\right) = .50$
Professor C. Normal	$P(at\ least\ 3\ mins)$ $\quad = 1 - P(less\ than\ 3\ mins.)$ $x = 3 \rightarrow z = \frac{3-3}{.60} = 0$ $P(X \geq 3) = 1 - \Phi(0)$ $\quad\quad = 1 - .5000 = .5000$
Professor C. Exponential	$P(at\ least\ 3\ mins)$ $\quad = 1 - P(less\ than\ 3\ mins.)$ $F(x; .50) = 1 - e^{-.50x}$ $F(3; .50) = 1 - e^{-.50 \cdot 3} = .7769$ $P(X \geq 3) = 1 - F(3)$ $\quad\quad = 1 - .7769 = .2231$

Continuous -- Probability of (X > 5)

Professor	What's the probability the overtime will exceed 5 minutes?
Professor C. Uniform	Not defined for $x > 5$
Professor C. Normal	$x = 5 \rightarrow z = \dfrac{5-3}{.60} = 3.3333$ $P(X \geq 5) = 1 - \Phi(3.333)$ $ = 1 - .9996 = .0004$
Professor C. Exponential	$P(X \geq 5) = 1 - P(X \leq 5)$ $ = 1 - P(x; .50)$ $ = 1 - (1 - e^{-.50x})$ $ = 1 - (1 - e^{-.50 \cdot 5}) \quad = 1 - (.9179) \; = .0821$

Continuous Distribution Summary

Probabilities of Overtime (mins./class)	Professor Uniform	Professor Normal	Professor Exponential
Expected Overtime	3.0	3.0	2.0
Probability of $(X = 3)$	0	0	0
Probability of $(2 \leq X \leq 4)$	.50	.905	.233
Probability of $(3 \leq X)$	.50	.50	.223

To minimize staying overtime for a class, for continuous distributions, take Professor Exponential's class.

Normal Distribution -- Additional Information

Find Percentile for Given Overtime

Question	Answer
If Professor Normal held the class late by 4.25 minutes, what percentile would this be?	First, convert the minutes to a Z score so we can use the Standard Normal Tables. Z is just the number of standard deviations from the mean. Here, $\mu = 3$ so we would expect the percentile to be above .50 (or 50%). $$z = \frac{x - \mu}{\sigma} = \frac{x - 3}{.60}$$ $$z = \frac{4.25 - 3}{.60} = 2.08$$ $z = 2.08$ ➔ Percentile = .9812 This means that 98% of the time, Professor Normal runs ≤ 4.25 minutes overtime. Also, for 2% of his classes, he runs more than 4.25 minutes overtime.

Find Overtime for Given Percentile

Question	Answer
Regarding Professor Normal, we know that the 50th percentile is at $\mu = 3$ minutes. That means that 50% of the time this professor will run late ≤ 3 minutes. What is the 33rd percentile for Professor Normal?	We are trying to find the overtime minutes associated with the 33trd percentile. Since this is less than the 50% we expect the number to be less than $\mu = 3$. Use the Standard Normal Table. • Look at the numbers in the middle of the chart for ".33" and work backwards to find the z-value associated with the 33rd percentile. • $z = -.44$ Now, convert the z-value to minutes. $$z = \frac{x - \mu}{\sigma} = \frac{x - 3}{.60}$$ $x = (.60)z + 3$ $x = (.60)(-.44) + 3 = 2.74$ minutes This means that 33% of the time, Professor Normal runs ≤ 2.74 minutes overtime. Also, for 67% of his classes, he runs more than 2.74 minutes overtime.

Compare Actual Data With Normal Distribution

Question
Throughout the term, this student randomly collected eight data points for Professor Normal's class.   He wrote down the number of minutes of overtime for 8 classes. Here is his data: 3.1, 3,2, 1.5, 4.4, 3.6, 1.2, 0.3, 2.5   Do the overtime minutes for Professor Normal's class actually follow a normal distribution?   In other words, should he change his name?

Solution: Create a probability plot to determine if the sample data matches a normal distribution.

- Calculate the percentile for n samples.
- Use standard normal table to find the associated standard z-percentiles for each percentile.
- Organize data samples, lowest to highest.
- Plot the (x, y) pairs of data with $x = z\_percentile$ and $y = sample\ observation$

$$percentage = \left(\frac{100}{n}\right)(i - .5)$$

$$percentile = \left(\frac{1}{n}\right)(i - .5)$$

Here: $n = 8$ and $i = 1, 2, 3 \ldots 8$

percentile	Z_percentile (x)	observed value (y)
.0625	−1.53	0.3
.1875	−0.89	1.2
.3125	−0.49	1.5
.4375	−0.16	2.5
.5625	0.16	3.1
.6875	0.49	3.2
.8125	0.89	3.6
.9375	1.53	4.4

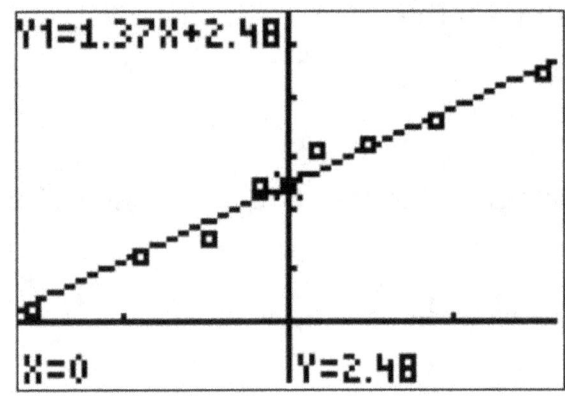

Normal Probability Plot

$x = z\_percentile$

$y = observed\ value$

The data points follow a reasonably straight line so it is plausible that the data has a Normal distribution.

Good News! There is no need for Professor Normal to change his name!

Create a Normal Probability Plot

Question
Throughout the term, this student randomly collected eight data points for Professor Normal's class. He wrote down the number of minutes of overtime for 8 classes. Here is his data: 3.1, 3,2, 1.5, 4.4, 3.6, 1.2, 0.3, 2.5 Use Minitab to create a Normal Probability Plot, using the random data samples collected for Professor Normal's class.

Note: The plots were created with the Minitab statistical utility. Other statistical software utilities are similar.

Do This	Professor Normal's Probability Plots
Minitab Probability Plot (easy default): • Enter your data in one column. • Graph/ Probability Plot/ Single • Select your data column • Select Distribution = Normal • Select Data Display = Both symbols & fit • Do not show confidence interval Note: The default Minitab Probability Plot has the data on the x-axis and the z-percentile on the y-axis.	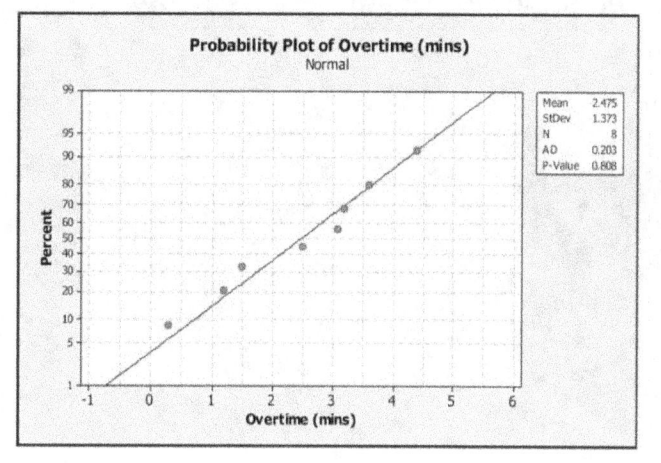

Do This	Professor Normal's Probability Plots
To calculate z-percentile directly from data • Enter data in C1 and name it "data" • Name column C2 "z-percentile" • Select column C2 (highlight column) • Editor/ Formulas/ Assign Formula to C2 • Use NSCORE(C1) Note: Minitab Scatterplot with Regression. Select z-percentile for X data and data observations for Y data.	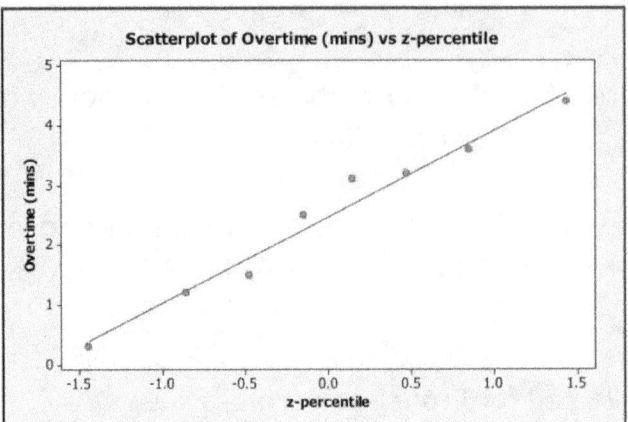

Create an "Ab-Normal" Probability Plot

Question
Professor Normal has a daughter named Abby who also teaches a class at this university. Does her overtime minutes follow a normal distribution? Another student in her class randomly collected the following six data points: 0.0, 0.25, 1.0, 4.0, 9.0, 15.0

Do This	Abby Normal's Probability Plots
Create a Minitab default Probability Plot to quickly check the data: The default Minitab Probability Plot for Abby Normal's overtime data distribution. Note: Abby Normal's data distribution appears "abnormal!"	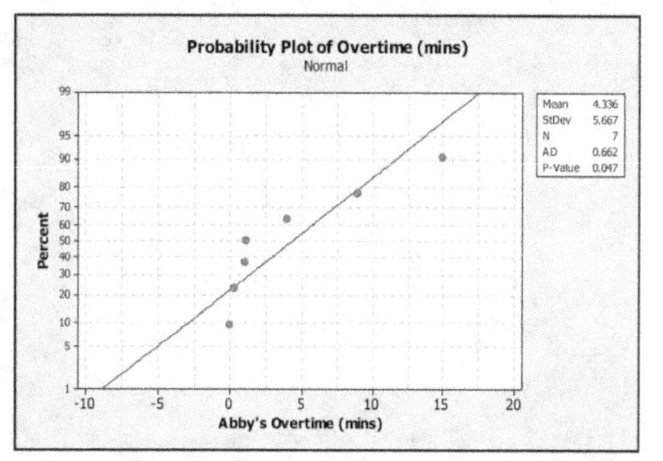
It does not appear to follow a normal distribution. Take the square root of each data sample and try again: Note: The square root of the data points follows a reasonably straight line so it is plausible that the square root of the data has a Normal distribution.	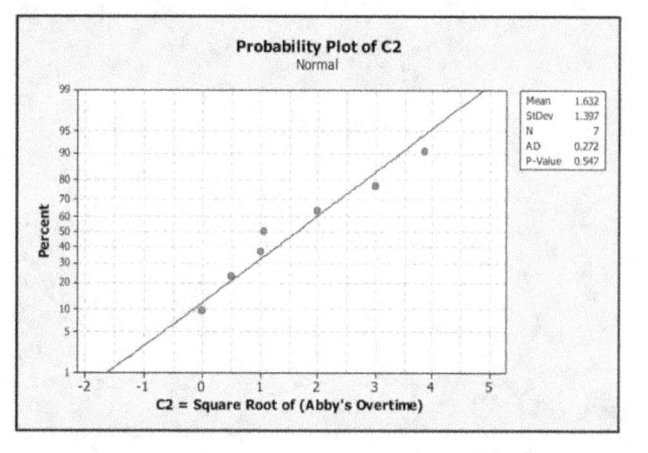

More Continuous Probability Distributions

Additional Continuous Distributions	• Standard Gamma
	• Chi-Squared (χ^2)

Background Information – Gamma Distribution

Gamma Distribution: $f(x; \alpha, \beta) = \frac{1}{\beta^\alpha \cdot \Gamma(\alpha)} \cdot x^{\alpha-1} e^{-\frac{x}{\beta}} \qquad x \geq 0$

The Gamma Function is: $\Gamma(\alpha) = \int_0^\infty x^{\alpha-1} e^{-x} dx$

- $\Gamma(\alpha) = (\alpha - 1) \cdot \Gamma(\alpha - 1) \qquad \alpha > 1$
- $\Gamma(n) = (n - 1)! \qquad n > 0$
- $\Gamma\left(\frac{1}{2}\right) = \sqrt{\pi}$

The Standard Gamma distribution is the Gamma distribution with $\beta = 1$

Standard Gamma Distribution:
$(x; \alpha) = \frac{x^{\alpha-1} e^{-x}}{\Gamma(\alpha)}$

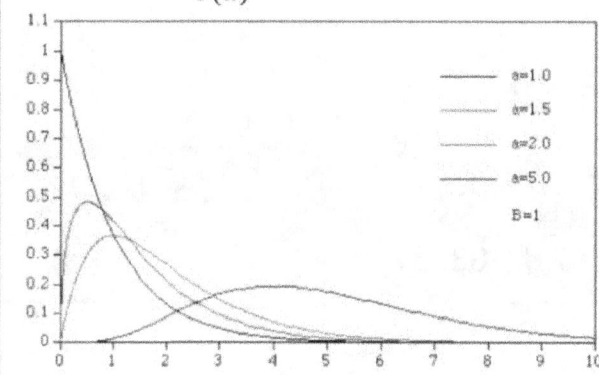

If $\alpha = 1$
It becomes an exponential distribution.

The Chi-Squared distribution is the Gamma distribution with $\alpha = v/2$ and $\beta = 2$

Chi-Squared Distribution:

$f(x; v) = \frac{1}{2^{v/2} \Gamma\left(\frac{v}{2}\right)} x^{\left(\frac{v}{2}\right)-1} e^{-x/2}$

$v =$ "degrees of freedom"

More Continuous Probability Distributions (pmf)

Meet the professors who represent the additional continuous distributions. Their middle name is "Continuous." They all hold their classes overtime, in their own way.

Professor	Probability Distributions (pmf)
Professor C. Standard Gamma	For this professor, let $\alpha = 3$. Then: $\Gamma(\alpha) = \Gamma(3) = 2! = 2$ $f(x) = \dfrac{x^2 \cdot e^{-x}}{2} \qquad x \geq 0 \qquad \beta = 1, \ \alpha = 3$
Professor C. Chi-Squared His friends call him χ^2	For this Professor, let $v = 8$ (degrees of freedom) Then: $\Gamma\left(\dfrac{v}{2}\right) = \Gamma(4) = (4-1)! = 6$ $f(x) = \dfrac{1}{2^4 \cdot 6} x^3 e^{-x/2} = \dfrac{1}{96} x^3 e^{-x/2}$ $b(x; n, p) = \binom{n}{x} p^x (1-p)^{n-x}$ $b\left(x; 5, \dfrac{1}{6}\right) = \binom{5}{x} (.1667)^x (.8333)^{5-x}$

Cumulative Probability (cdf)

Professor	Cumulative Probability (cdf)
Professor C. Standard Gamma	$F(x, \alpha) = \int_0^x \frac{y^{\alpha-1} e^{-y}}{\Gamma(\alpha)} dy$ or $F(x, \alpha, \beta) = F\left(\frac{x}{\beta}, \alpha\right)$ Tables Easier to use Gamma Function table Here: $\beta = 1$ and $\alpha = 3$ and $\Gamma(3) = (3-1)! = 2$ $F(x) = \int_0^x \frac{y^2 e^{-y}}{2} dy = \frac{1}{2} \int_0^x (y^2 e^{-y}) dy$ $\beta = 1, \alpha = 3$
Professor C. Chi-Squared (χ^2)	Here: $v =$ degrees of freedom $= 8$ Here: $\alpha = \frac{v}{2} = 4$ and $\beta = 2$ $F(x) = \int_0^x \frac{1}{96} y^3 e^{-y/2} dy \qquad v = 8$ and $\beta = 2$

Expected Values and Basic Statistics

Professor	What value would you expect?	What are the basic statistics? (mean, variance, and standard deviation)
Professor C. Standard Gamma	$E(X) = \mu = \alpha\beta$ $E(X) = \alpha\beta$ $E(X) = (3)(1) = 3$	$mean = \mu = E(X) = 3$ $V(X) = \alpha\beta^2$ $V(X) = (3)(1) = 3$ $\sigma = \sqrt{V(X)} = \sqrt{3} = 1.732$
Professor C. Chi-Squared (χ^2)	$E(X) = \mu = \alpha\beta$ For Chi-Squared: $\alpha = v/2$ and $\beta = 2$ Here: $v = 8$ ➔ $\alpha = 4$ $E(X) = \mu = \alpha\beta$ $E(X) = (4)(2) = 8$	$mean = \mu = E(X) = 8$ $V(X) = \alpha\beta^2$ $V(X) = (4)(2)^2 = 16$ $\sigma = \sqrt{V(X)} = \sqrt{16} = 4$

Probability of (X = 3)

Professors	Probability of staying exactly 3 minutes overtime.
Professor C. Standard Gamma	$P(X = 3) = ???$ Can NOT calculate probabilities for a single value. Can only calculate probabilities for a range of x.
Professor C. Chi-Squared (χ^2)	$P(X = 3) = ???$ Can NOT calculate probabilities for a single value. Can only calculate probabilities for a range of x.

Probability of (2 <= X <= 4)

Professor	What's the probability of staying between 2 and 4 min. Inclusive?
Professor C. Standard Gamma	$P(2 \leq X \leq 4) = F(4) - F(2)$ $= \frac{1}{2}\int_0^4 (y^2 e^{-y})\, dy - \frac{1}{2}\int_0^2 (y^2 e^{-y})\, dy$ $= \frac{1}{2}[(2 - 26e^{-4}) - (2 - 10e^{-2})]$ $= -13e^{-4} + 5e^{-2} = .4386$ Using Gamma Tables for: $\beta = 1,\ \alpha = 3$ $P(2 \leq X \leq 4) = F\left(\frac{4}{1}, 3\right) - F\left(\frac{2}{1}, 3\right)$ $= F(4, 3) - F(2, 3) = .762 - .323 = .439$ Note: Easier to use tables!

Professor	What's the probability of staying between 2 and 4 min. Inclusive?
Professor C. Chi-Squared (χ^2)	$P(2 \leq X \leq 4) = F(4) - F(2)$ $= \frac{1}{96} \int_0^4 y^3 e^{-y/2} \, dy - \frac{1}{96} \int_0^2 y^3 e^{-y/2} \, dy$ $= \frac{1}{96}[\,13.716 - 1.823\,] = \frac{1}{96}[\,11.893\,] = .12389$ Using Standard Chi-Squared Tables for: $v = 8$ Note: Use these tables differently. Since tables show "upper tail" area, subtract it from 1 to get "lower tail" area which is the cumulative probability. Also, the "x" is in the middle of the table and you must interpolate to get an accurate area (α). $P(2 \leq X \leq 4) = F(4) - F(2)$ $= (1 - .8586) - (1 - .9800)$ $= -.8586 + .9800 = .1214$ $\frac{.10 - .90}{13.312 - 3.490} = \frac{n - .90}{4 - 3.490}$ → $n = .85867$ $\frac{.975 - .99}{2.180 - 1.646} = \frac{n - .99}{2 - 1.646}$ → $n = .9800$ Note: For Chi-Squared distributions, integration is easier than using the tables!

Incentives

Money Incentives - Description

The student's family (Mother, Father, Aunt, and Uncle) want to encourage the student to stay in class during overtime. So, each person offers their own incentive (reward) to encourage the student to stay overtime.

Person giving the incentive	Incentive Description
Mother	The student's mother will pay him a quarter ($0.25) for every minute of overtime.
Father	The student's father will pay him $10.00 for any class that runs 5 or more minutes late.
Aunt	The student's aunt will pay him $5.00 per minute for any class that runs overtime, up to 3 minutes.
Uncle	The student's uncle will pay him $20.00 per minute for any class that runs overtime by 4 or more minutes.

Money Incentive Calculations – Discrete

Discrete Incentive Calculations – Mom

Professor	Discrete Distributions Calculations for Mom's Incentive $h(x) = payment\ function = .25x$	Mom's total incentive For 30 Classes
Professor D. Discrete	$E[h(x)] = \sum h(x) \cdot p(x)$ $E[h(x)] = h(\mu) = .25(2.21) = .5525$ Variance $= E[h^2(x)] - E[h(x)]^2$ Variance $= (slope)^2 \cdot V(X)$ Variance $= (.25)^2 \cdot (2.366) = .1479$ With: $\sigma = \sqrt{.1479} = .3846$	Mom's total $= 30(.5525)$ $= \$16.58$
Professor D. Binomial	$E[h(x)] = \sum h(x) \cdot p(x)$ $E[h(x)] = h(\mu) = .25(.8333) = .2083$ Variance $= E[h^2(x)] - E[h(x)]^2$ Variance $= (slope)^2 \cdot V(X)$ Variance $= (.25)^2 \cdot (.6945) = .0434$ With: $\sigma = \sqrt{.0434} = .2083$	Mom's total $= 30(.2083)$ $= \$6.25$

Mom's Incentive Calculations Continued ...

Professor	Discrete Distributions Calculations for Mom's Incentive $h(x) = payment\ function = .25x$	Mom's total incentive For 30 Classes
Professor D. Poisson	$E[h(x)] = \sum h(x) \cdot p(x)$ $E[h(x)] = h(\mu) = .25(2) = .50$ Variance $= E[h^2(x)] - E[h(x)]^2$ Variance $= (slope)^2 \cdot V(X)$ Variance $= (.25)^2 \cdot (2) = .125$ With: $\sigma = \sqrt{.125} = .3536$	Mom's total $= 30(.50)$ $= \$15.00$
Professor D. Geometric	$E[h(x)] = \sum h(x) \cdot p(x)$ $E[h(x)] = h(\mu) = .25(5) = 1.25$ Variance $= E[h^2(x)] - E[h(x)]^2$ Variance $= (slope)^2 \cdot V(X)$ Variance $= (.25)^2 \cdot (30) = 1.88$ With: $\sigma = \sqrt{1.88} = 1.34$	Mom's total $= 30(1.25)$ $= \$37.70$

Discrete Incentive Calculations – Dad

Professor	Discrete Distributions: Expected Incentive Calculations for Dad's Incentive $h(x) = 10 \quad for \ x \geq 5$	Dad's total incentive For 30 Classes
Professor D. Discrete	$E[h(x)] = \sum h(x) \cdot p(x)$ $= (10) \cdot P(X \geq 5) = (10)[1 - P(X \leq 4)]$ $= (10)[1 - .85] = (10)[.15] = \$1.50$ $E[h^2(x)] = \sum h^2(x) \cdot p(x)$ $= (10^2) \cdot P(X \geq 5) = (10^2)[1 - P(X \leq 4)]$ $= (10^2)[1 - .85] = 15$ NOTE: Variance $\neq (slope)^2 \cdot V(X)$ Variance $= E[h^2(x)] - E[h(x)]^2$ Variance $= 15 - [1.5]^2 = 12.75$ With: $\sigma = \sqrt{12.75} = 3.57$	Dad's total $= 30(1.50)$ $= \$45.00$
Professor D. Binomial	$E[h(x)] = \sum h(x) \cdot p(x)$ $= (10) \cdot P(X \geq 5) = (10)[1 - P(X \leq 4)]$ $= (10)[1 - B(4; 5, .17)] = (10)[1 - 1]$ $= (10)[0] = \$0$	Dad's total $= 30(0)$ $= \$0$

Dad's Incentives Calculations Continued ...

Professor	Discrete Distributions: Expected Incentive Calculations for Dad's Incentive $h(x) = 10 \quad for\ x \geq 5$	Dad's total incentive For 30 Classes
Professor D. Poisson	$E[h(x)] = \sum h(x) \cdot p(x)$ $= (10) \cdot P(X \geq 5) = (10)[1 - P(X \leq 4)]$ $= (10)[1 - F(4; 2)] = (10)[1 - .947]$ $= (10)[0.053] = \$0.53$	Dad's total $= 30(.53)$ $= \$15.90$
Professor D. Geometric	$E[h(x)] = (10) \cdot \{P(X \geq 5)\}$ $= (10) \cdot \left\{\left(\frac{5}{6}\right)^5\right\} = (10)\{.402\} = 4.02$	Dad's total $= 30(4.02)$ $= \$120.60$
	Another Calculation (inefficient) $= (10)[1 - P(X \leq 4)]$ $= (10) \cdot$ $\quad \{1 - [p(4) + p(3) + p(2) + p(1) + p(0)]\}$ $=$ $(10)\left\{1 - \left[\begin{array}{c}\left(\frac{5}{6}\right)^4\left(\frac{1}{6}\right) + \left(\frac{5}{6}\right)^3\left(\frac{1}{6}\right) + \left(\frac{5}{6}\right)^2\left(\frac{1}{6}\right) \\ + \left(\frac{5}{6}\right)^1\left(\frac{1}{6}\right) + \left(\frac{5}{6}\right)^0\left(\frac{1}{6}\right)\end{array}\right]\right\}$ $= 10\{1 - [.598]\} = 10\{.402\} = \$4.02$	

Discrete Incentive Calculations – Aunt

Professor	Discrete Distributions: Expected Incentive $h(x) = 5x \quad 0 \leq x \geq 3$	Aunt's total incentive For 30 Classes
Professor D. Discrete	$E[h(x)] = \sum_1^3 5x \cdot p(x)$ $= 5(1)(.50) + 5(2)(.10) + 5(3)(.20)$ $= 2.5 + 1 + 3 = \$6.50$	Aunt's total $= 30(6.50)$ $= \$195.00$
Professor D. Binomial	$E[h(x)] = \sum_1^3 5x \cdot p(x)$ $= 5(1) \cdot b(1; 5, .17) + 5(2) \cdot b(2; 5, .17)$ $\qquad + 5(3) \cdot b(3; 5, .17)$ $= 5 \binom{5}{1}(.17)^1(.83)^4 + 10 \binom{5}{2}(.17)^2(.83)^3$ $\qquad + 15 \binom{5}{3}(.17)^3(.83)^2$ $= 2.017 + 1.635 + .5077 = \$4.18$	Aunt's total $= 30(4.18)$ $= \$125.31$
Professor D. Poisson	$E[h(x)] = \sum_1^3 5x \cdot p(x)$ $= 5(1)p(1; 2) + 5(2)p(2; 2) + 5(3)p(3; 2)$ $= 5 \cdot \frac{e^{-2} 2^1}{1!} + 10 \cdot \frac{e^{-2} 2^2}{2!} + 15 \cdot \frac{e^{-2} 2^3}{3!}$ $= 1.3535 + 2.707 + 2.707 = \$6.77$	Aunt's total $= 30(6.77)$ $= \$203.10$
Professor D. Geometric	$E[h(x)] = \sum_1^3 5x \cdot p(x)$ $= 5(1) \cdot p(1) + 5(2) \cdot p(2) + 5(3) \cdot p(3)$ $= 5 \left(\frac{5}{6}\right)^1 \left(\frac{1}{6}\right) + 10 \left(\frac{5}{6}\right)^2 \left(\frac{1}{6}\right) + 15 \left(\frac{5}{6}\right)^3 \left(\frac{1}{6}\right)$ $= 5(.139) + 10(.116) + 15(.096) = \$3.30$	Aunt's total $= 30(3.30)$ $= \$99.00$

Discrete Incentive Calculations – Uncle

Professor	Discrete Distributions: Expected Incentive Calculations for Uncle's Incentive $h(x) = 20x \qquad 4 \leq x$	Uncle's total incentive For 30 Classes
Professor D. Discrete	$E[h(x)] = \sum_{4}^{5} 20x \cdot p(x)$ $= 20(4)(.05) + 20(5)(.15)$ $= 4 + 15 = \$19.00$	Uncle's total $= 30(19.00)$ $= \$570.00$
Professor D. Binomial	$E[h(x)] = \sum_{4}^{5} 20x \cdot p(x)$ $= 20(4) \cdot b(4; 5, .17) + 20(5) \cdot b(5; 5, .17)$ $= 80 \binom{5}{4}(.17)^4(.83)^1 + 100 \binom{5}{5}(.17)^5(.83)^0$ $= .277 + .0142 = .291$	Uncle's total $= 30(.291)$ $= \$8.73$
Professor D. Poisson	$E[h(x)] = \sum_{4}^{5} 20x \cdot p(x)$ $= 20(4) \cdot p(4; 2) + 20(5) \cdot p(5; 2)$ $= 80 \cdot \frac{e^{-2} 2^4}{4!} + 100 \cdot \frac{e^{-2} 2^5}{5!}$ $= 7.22 + 3.61 = 10.83$	Uncle's total $= 30(10.83)$ $= \$324.90$

Uncle's Incentives Calculations Continued ...

Professor	Discrete Distributions: Expected Incentive Calculations for Uncle's Incentive $h(x) = 20x \quad 4 \leq x$	Uncle's total incentive For 30 Classes
Professor D. Geometric	$E[h(x)] = \sum_{4}^{5} 20x \cdot p(x)$ Theoretically, this professor may have an infinite number of failures before success! But, luckily, the probability of an infinite number of failures is almost zero! Remember, we are adding money. So, let's just agree to stop adding when the numbers (additional money) gets very small (less than one penny). $E[h(x)] = 20(4) \cdot p(4) + 20(5) \cdot p(5) +$ $\qquad + 20(6) \cdot p(6) + 20(7) \cdot p(7) + \ldots$ Recall: $p(x) = (1-p)^x p = \left(\frac{5}{6}\right)^x \left(\frac{1}{6}\right)$ $E[h(x)] = 20(4)(.0804) + 20(5)(.0667) +$ $\qquad + 20(6)(.0558) + 20(7)(.0465) +$ $\qquad + 20(8)(.0389) + \cdots$ Easier to use a calculator. See next page ...	SEE NEXT PAGE

Uncle's Incentives Calculations Continued ...

Professor	Discrete Distributions: Expected Incentive Calculations for Uncle's Incentive $h(x) = 20x \qquad 4 \leq x$	Uncle's total incentive For 30 Classes
Professor D. Geometric	Using the summation function on a calculator ... $E[h(x)] = \sum_{4}^{\infty} 20x \cdot p(x)$ $E[h(x)] = \sum_{4}^{\infty} 20x \cdot \left[\left(\frac{5}{6}\right)^x \left(\frac{1}{6}\right)\right]$ $E[h(x)] = \left(\frac{20}{6}\right) \sum_{4}^{\infty} 20x \cdot \left(\frac{5}{6}\right)^x$ Try some upper limits: $\left(\frac{20}{6}\right) \sum_{4}^{30} x \cdot \left(\frac{5}{6}\right)^x = 84.28$ $\left(\frac{20}{6}\right) \sum_{4}^{40} x \cdot \left(\frac{5}{6}\right)^x = 86.28$ $\left(\frac{20}{6}\right) \sum_{4}^{50} x \cdot \left(\frac{5}{6}\right)^x = 86.70$ $\left(\frac{20}{6}\right) \sum_{4}^{60} x \cdot \left(\frac{5}{6}\right)^x = 86.79$ $\left(\frac{20}{6}\right) \sum_{4}^{70} x \cdot \left(\frac{5}{6}\right)^x = 86.80$ $\left(\frac{20}{6}\right) \sum_{4}^{80} x \cdot \left(\frac{5}{6}\right)^x = 86.80$ Therefore: $E[h(x)] = \sum_{4}^{\infty} 20x \cdot p(x) \approx \$86.80$	Uncle's total = 30(86.80) = $2604.00

Discrete Incentive -- Summary

Total Incentive ($) For 30 Classes	Professor D. Discrete	Professor D. Binomial	Professor D. Poisson	Professor D. Geometric
Mom	16.58	6.25	15.00	37.50
Dad	45.00	0	15.90	120.60
Aunt	195.00	125.31	203.10	99.00
Uncle	570.00	8.73	324.90	2604.00
TOTAL	826.58	140.29	558.9	2861.1

To maximize incentive money from relatives, for discrete distributions, the student should take Professor D. Geometric's class.

Money Incentive Calculations – Continuous

Continuous Incentive Calculations – Mom

Professor	Continuous Distributions Calculations for Mom's Incentive $h(x) = payment\ function = .25x$	Total Mom Incentive For 30 Classes
Professor C. Uniform	$E[h(x)] = h(\mu)$ Expected payment $= h(\mu) = .25(3) = .75$ Variance $= (slope)^2 \cdot V(X)$ $\qquad = (.25)^2 \cdot (1.333) = .08331$ $\sigma = \sqrt{.08331} = .2886$	Mom's total $= 30(.75)$ $= \$22.58$
Professor C. Normal	$E[h(x)] = \sum h(x) \cdot p(x)$ $E[h(x)] = h(\mu)$ Expected payment $= h(\mu) = .25(3) = .75$ Variance $= (slope)^2 \cdot V(X)$ $\qquad = (.25)^2 \cdot (1.333) = .08331$ $\sigma = \sqrt{.08331} = .2886$	Mom's total $= 30(.75)$ $= \$22.58$
Professor C. Exponential	$E[h(x)] = \sum h(x) \cdot p(x)$ $E[h(x)] = h(\mu)$ Expected payment $= h(\mu) = .25(2) = .50$ Variance $= (slope)^2 \cdot V(X)$ $\qquad = (.25)^2 \cdot (.25) = .1563$ $\sigma = \sqrt{.1563} = .125$	Mom's total $= 30(.50)$ $= \$15.00$

Continuous Incentive Calculations – Dad

Professor	Continuous Distributions Calculations for Dad's Incentive $h(x) = 10 \quad for \ x \geq 5$	Total Dad Incentive For 30 Classes
Professor C. Uniform	$E[h(x)] = \sum h(x) \cdot p(x)$ $= (10) \cdot P(X \geq 5)$ $= (10) \cdot [1 - P(X \leq 5)]$ $= (10)[1-1] = (10)[0] = 0$	Dad's total $= 30(0)$ $= \$0$
Professor C. Normal	$x = 5 \quad \rightarrow \quad z = \frac{5-3}{.60} = 3.333$ $E[h(x)] = \sum h(x) \cdot p(x)$ $= (10) \cdot P(X \geq 5)$ $= (10) \cdot [1 - P(X \leq 5)]$ $= (10) \cdot [1 - \Phi(3.333)]$ $= (10) \cdot [1 - .9996] = .004$	Dad's total $= 30(.004)$ $= \$0.12$
Professor C. Exponential	$F(5;.50) = 1 - e^{-50 \cdot .5} = .9179$ $E[h(x)] = \sum h(x) \cdot p(x)$ $= (10) \cdot P(X \geq 5)$ $= (10) \cdot [1 - P(X \leq 5)]$ $= (10) \cdot [1 - (5;.50)]$ $= (10) \cdot [1 - .9179]$ $= (10) \cdot [.0821] = .821$	Dad's total $= 30(.821)$ $= \$24.63$

Continuous Incentive Calculations – Aunt

Professor	Continuous Distributions Calculations for Aunt's Incentive $h(x) = 5(x)$ $\quad 0 \leq x \geq 3$	Total Aunt Incentive For 30 Classes
Professor C. Uniform	$E[h(x)] = \int_1^3 5x \cdot p(x)\, dx$ $= \int_1^3 5x \cdot \left(\frac{1}{4}\right) dx = \left(\frac{5}{4}\right) \int_1^3 x\, dx$ $= \frac{5}{4} \left[\frac{x^2}{2}\right]_1^3 = \frac{5}{8}[9-1] = 5$	Aunt's total $= 30(5.00)$ $= \$150.00$
Professor C. Normal	$E[h(x)] = \int_0^3 5x \cdot p(x)\, dx$ $= \int_0^3 5x \cdot \left(\frac{1}{\sqrt{2\pi}} e^{-\frac{z^2}{2}}\right) dx$ $= \left(\frac{5}{\sqrt{2\pi}}\right) \int_0^3 x \cdot e^{\left(\frac{-x^2 + 6x - 9}{.72}\right)} dx$ $= \frac{5}{\sqrt{2\pi}} [1.896] = 1.509$	Aunt's total $= 30(1.509)$ $= \$45.27$
	$z = \frac{x-3}{.60}$ $-\frac{z^2}{2} = \frac{-(x-3)^2}{.72} = \frac{-x^2 + 6x - 9}{.72}$	
Professor C. Exponential	$E[h(x)] = \int_0^3 5x \cdot p(x)\, dx$ $= \int_0^3 5x \cdot f(x; .50)\, dx$ $= \int_0^3 5x \cdot (.50 \cdot e^{-.50x})\, dx$ $= 2.50 \int_0^3 x \cdot e^{-.50x}\, dx$ $= 2.50\,[1.769] = 4.422$	Aunt's total $= 30(4.42)$ $= \$132.60$

Continuous Incentive Calculations – Uncle

Professor	Continuous Distributions Calculations for Uncle's Incentive $h(x) = 20(x) \qquad 4 \leq x$	Total Uncle Incentive For 30 Classes
Professor C. Uniform	$E[\,h(x)\,] = \int_4^5 20x \cdot p(x)\, dx$ $= \int_4^5 20x \cdot \left(\frac{1}{4}\right) dx = (5) \int_4^5 x\, dx$ $= \frac{5}{2}\,[\,x^2\,]_4^5 = 2.50\,[\,25 - 16\,] = 22.5$	Uncle's total $= 30(22.50)$ $= \$675.00$
Professor C. Normal	$E[\,h(x)\,] = \int_4^\infty 20x \cdot p(x)\, dx$ $= \int_4^\infty 20x \cdot \left(\frac{1}{\sqrt{2\pi}}\, e^{-\frac{z^2}{2}}\right) dx$ $= \left(\frac{20}{\sqrt{2\pi}}\right) \int_4^\infty x \cdot e^{\left(\frac{-x^2 + 6x - 9}{.72}\right)} dx$ $= \frac{20}{\sqrt{2\pi}}\,[\,.3054\,] = .9721$ $z = \frac{x-3}{.60}$ $-\frac{z^2}{2} = \frac{-(x-3)^2}{.72} = \frac{-x^2 + 6x - 9}{.72}$	Uncle's total $= 30(.9721)$ $= \$29.16$
Professor C. Exponential Never runs over by more than 5 mins.	$E[\,h(x)\,] = \int_4^\infty 20x \cdot p(x)\, dx$ $= \int_4^\infty 20x \cdot f(x;\,.50)\, dx$ $= \int_4^\infty 20x \cdot (.50 \cdot e^{-.50x})\, dx$ $= 10 \int_4^5 x \cdot e^{-.50x}\, dx$ $= 10\,[\,1.642\,] = 16.24$	Uncle's total $= 30(16.24)$ $= \$487.20$

Continuous Incentive -- Summary

Total Incentive ($) For 30 Classes	Professor C. Uniform	Professor C. Normal	Professor C. Exponential
Mom	22.58	22.58	15.00
Dad	0	0.12	24.63
Aunt	150.00	45.27	132.60
Uncle	675.00	29.16	487.20
TOTAL	847.58	97.13	659.43

To maximize money-incentives from relatives, for continuous distributions, the student should take Professor C. Uniform's class.

Appendix

The Original Problem

This problem was the inspiration for this detailed comparison of discrete and continuous distributions.

Problem Statement

A college professor never finishes his lecture before the end of the hour and always finishes his lectures within 2 min after the hour. Let X = the time that elapses between the end of the hour and the end of the lecture and suppose the PDF of X is:

$$f(x) = \begin{cases} kx^2, & 0 \le x \le 2 \\ 0, & otherwise \end{cases}$$

a) Find the value of k and draw the corresponding density curve. (Hint: Total area under the graph of f(x) is 1.)
b) What is the probability that the lecture ends within 1 min of the end of the hour?
c) What is the probability that the lecture continues beyond the hour for between 60 and 90 sec?
d) What is the probability that the lecture continues for at least 90 sec beyond the end of the hour?

Source: Probability & Statistics for Engineering and the Sciences, Jay L. Devore, Eighth Edition,
Chapter 4, Continuous Random Variables & Probability Distributions, Page 142, Question #5.

Problem Solution

#	Question	Answer
a.	Find the value of k and draw the corresponding density curve Hint: Total area under graph of $f(x)$ is 1. Recall: $f(x) = \begin{cases} kx^2, & 0 \leq x \leq 2 \\ 0, & \text{otherwise} \end{cases}$	Area under curve = 1. $\int_0^2 kx^2\, dx = 1$ $k \int_0^2 x^2\, dx = 1$ $k \left[\frac{1}{3} x^3 \right]_0^2 = 1$ $\frac{k}{3}[8 - 0] = 1$ $\left(\frac{8}{3}\right) k = 1$ $k = \frac{3}{8}$
b.	What is the probability that the lecture ends within 1 min of the end of the hour?	$P(X \leq 1) = \int_0^1 f(x)\, dx = \int_0^1 kx^2\, dx$ $= \left(\frac{3}{8}\right) \int_0^1 x^2\, dx$ $= \left(\frac{3}{8}\right) \left[\frac{1}{3} x^3 \right]_0^1$ $= \left(\frac{1}{8}\right) [1^3 - 0]$ $= \frac{1}{8} = .422$

#	Question	Answer
c.	What is the probability that the lecture continues beyond the hour for between 60 and 90 seconds?	$P(1 \leq X \leq 1.5) =$ $= \int_1^{1.5} \left(\frac{3}{8}\right) x^2 \, dx$ $= \left(\frac{3}{8}\right) \int_1^{1.5} x^2 \, dx$ $= \left(\frac{3}{8}\right) \left[\frac{1}{3} x^3\right]_1^{1.5}$ $= \left(\frac{1}{8}\right) [1.5^3 - 1^3]$ $= \left(\frac{1}{8}\right) [3.375 - 1] = .297$
d.	What is the probability that the lecture continues for at least 90 seconds beyond the end of the hour?	$P(X \geq 1.5) =$ $= 1 - P(X \leq 1.5)$ $= 1 - \int_0^{1.5} \left(\frac{3}{8}\right) x^2 \, dx$ $= 1 - \left(\frac{3}{8}\right) \left[\frac{1}{3} x^3\right]_0^{1.5}$ $= 1 - \left(\frac{1}{8}\right) [1.5^3 - 0]$ $= 1 - \left(\frac{1}{8}\right) [3.375]$ $= .578$

The Optimist's Creed

For Professor D. Geometric …

Source: www.theparisreview.org

Other Books by Kathryn Paulk

Other Books by Kathryn Paulk

- Algebra 1 Help
- Algebra 2 Help
- Pre-Calculus and Trig Help
- College Algebra Help
- Pre-Calculus and Trig Problems & Solutions

- Calculus 1 Review in Bite-Size Pieces
- Calculus 2 Review in Bite-Size Pieces
- Calculus 3 Review in Bite-Size Pieces
- Differential Equations With Applications: Class Notes With Examples

- One-Page Summaries for Algebra, Geometry, and Pre-Calculus
- Graphing Functions Using Transformations for Algebra and Pre-Calculus
- Complex Numbers and Polar Curves For Pre-Calc and Trig: With Problems and Detailed Solutions
- Discrete and Continuous Probability Distributions: A Creative Comparison (V2)

- Teach Your Child to SWIM

BIG MATH For Little Kids

Workbooks & Solution Manuals for Young Children

- Introduction to Numbers (ages 2 – 5 yrs.)

- Introduction to Fractions
  by Sharing Things   (ages 3 – 8 yrs.)

- Introduction to Counting and Fractions
  by Cooking Breakfast (ages 5 – 15 yrs.)

- Learn About Fractions
  by Baking Cookies (ages 8 – 15 yrs.)

- Adding Big Numbers, Guessing Numbers
  and Secret Codes (ages 8 – 15 yrs.)

- Learn to Graph by Riding Bikes
  on Graph Paper   (ages 10 – 16 yrs.)

These books are based on the activities Kathy did
with her own son, when he was young.

www.ingramcontent.com/pod-product-compliance
Lightning Source LLC
Chambersburg PA
CBHW081500220526
45466CB00008B/2722